*To Harrison and Naila*

*Thank you for the inspiration*

# Table of Contents

# Introduction

There is a story from ancient Greece about one of Pythagoras's followers, a man called Hippasus. He insisted that the number, when multiplied by itself which resulted in two, couldn't be represented as a fraction.

This was almost heresy at the time because the general belief was that every natural mathematical relationship could be represented as a fraction. Legend has it that he was thrown off a cliff into the sea for daring to suggest such a thing. It was thought that natural things could be represented by a ratio of two whole numbers, which is another way of saying a fraction.

It turns out to be quite easy to prove that Hippasus was correct and it seems that the Pythagoreans soon realised this, so, red faces all round.

The word for one of these numbers is 'surd' and you may be interested to note that the word 'absurd' comes from the same latin word 'surdus' .

This book has the following structure:

- the format of a surd
- irrational numbers
- why use surds?
- methods for rationalising surds
- calculating the approximate value of surds

# Square Roots

First of all, what is a square root? It is a number which, when multiplied by itself results in the number that it is the square root of. For example 2 x 2 = 4 so 2 is a square root of 4.

-2 x -2 also equals 4 so -2 is also a square root of 4

Likewise, 3 and -3 are square roots of 9

For the rest of this book  we will simplify things by only using positive roots; these are known as the Principal Roots.

Roots can have an index – for example the 4$^{th}$ root of 81 is 3. because 81 = 3 x 3 x 3 x 3

The symbol we use to represent a square root is:

$\sqrt{\phantom{x}}$ it is called the radical symbol.

The number under the radical symbol is called the radicand; this is the number we are taking the root of. For example, for $\sqrt{5}$ , the radicand is 5 .

The number which indicates the power to which the root is raised to result in the radicand is called the index. If the index is not present, it is assumed to be 2.

radical
symbol

This example states that
the 3rd root of 8 is 2.        index

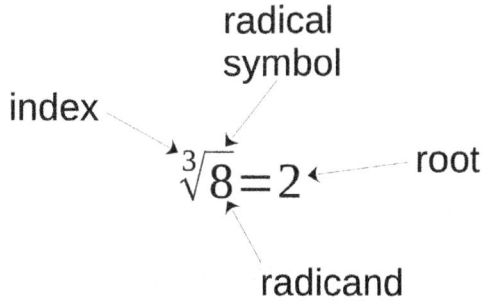

$$\sqrt[3]{8} = 2$$

root

radicand

# Surds – Absurd and Irrational

A surd is the root of a number, but not any number. For example, the square root of 4 is 2, and 2 is just a regular integer[1]. $\sqrt{4}$ is not a surd.

However, the square root of two is 1.41421356237309… the dots mean that it goes on forever. Additionally, it does not have any repeating groups at the end. The square root of 2 is a surd.

$$\sqrt{4}=2$$
$$\sqrt{4} \text{ is not a surd}$$
$$\sqrt{2}=1.41421356237309\dots$$
$$\sqrt{2} \text{ is a surd}$$

So what makes a surd a surd?

It is an irrational number. This means a number that can not be expressed as the fraction of two integers. It has a never-ending sequence of digits that also does not terminate with a repeating sequence.

For example $\dfrac{22}{7}$ as a decimal is 3.142857142857142857…

This can be expressed better as $3.\overline{142857}$ *(the line above the numbers indicates that they repeat continuously)*

The fact that $\dfrac{22}{7}$ is equivalent to a decimal with an infinitely repeating sequence at the end shows that the decimal equivalent is not a surd, even though it does not terminate.

If a never-ending number that is a surd is multiplied by itself, the result is the radicand of the surd.

---

1    Well 2 is not just a regular integer, but the first of the primes and also the only even prime, but that's another story

$$\sqrt{2} = 1.41421356237309\ldots$$
$$but$$
$$1.41421356237309\ldots \times 1.41421356237309\ldots = 2$$

The terminology for an irrational number is a number that does not terminate or have a recurring sequence at the end. However, if the infinite decimal series that is a surd is multiplied by itself, the result is a rational number

It should be noted that all irrational numbers are not necessarily surds. For example, $\pi$ is an irrational number, but so is $\pi^2$ so squaring the infinite series that makes up $\pi$ does not result in a rational number.

This implies that rational numbers do either terminate or recur. Therefore any number that either terminates or recurs can be represented as a fraction and is not a surd. Lets see how we can prove this.

## Terminating Decimals

A terminating decimal has a specific number of decimal places. To convert a terminating decimal to a fraction, the digits after the decimal point is the numerator of the fraction. The denominator is a one followed by the same number of zeros as there are digits in the numerator. For example:

$$0.3 = \frac{3}{10}$$

$$27.33 = 27\frac{33}{100}$$

$$0.573 = \frac{573}{1000}$$

No matter how many digits follow the decimal point, if it terminates it can still be converted to a fraction. Therefore terminating decimals are not surds.

## Recurring Decimals

Lets look a 1.3333... or we could write this as $1.\overline{3}$ It has an infinite number of 3s indicated by the line above the three so it is not a terminating number.

We want to convert $1.\overline{3}$ to a decimal.

First, let $x = 1.\overline{3}$

Now multiply both sides of this equation by 10 giving

$10x = 13.\overline{3}$ *(note that there are still an infinite number of threes after the decimal point)*

Now, subtract the first equation from the second:

$$10x = 13.\overline{3}$$
$$-\ x =\ \ 1.\overline{3}$$
$$9x = 12$$

$$x = \frac{12}{9} = \frac{4}{3}$$

So, $1.\overline{3}$ is equivalent to $\frac{4}{3}$ or $1\frac{1}{3}$ which is a rational number.

Lets try another, more complex number with a recurring decimal 23.4767676... or we can write this as $23.4\overline{76}$

Let x = 23.4$\overline{76}$ then multiply this equation by 10 giving

10x = 234.$\overline{76}$

Now multiply the original equation by 1000 giving

1000x = 23476.$\overline{76}$

What we are doing here is finding two multiples of x that are followed by the recurring part of the decimal. Now, get rid of the recurring decimals by subtracting the second equation from the third:

$$\begin{aligned} 1000x &= 23476.\overline{76} \\ -\quad 10x &= \phantom{234}234.\overline{76} \\ \hline 990x &= 23242 \end{aligned}$$

$$x = \frac{23242}{990} = \frac{11621}{495} = 23\frac{236}{495}$$

This method can be used for any number of recurring digits. Therefore, numbers that continue with recurring numbers or recurring sequences of numbers are rational numbers.

## Perfect Squares

A perfect square is the result of multiplying an integer by itself. It represents the area of a geometric square where the length of the sides is an integer.

The square root of all of the integers that are perfect squares are also integers and therefore not surds.

For example, in increasing order:

$$\sqrt{4}=2$$
$$\sqrt{9}=3$$
$$\sqrt{16}=4$$
$$\sqrt{25}=5$$
$$\sqrt{36}=6$$
$$\sqrt{49}=7$$
$$\sqrt{64}=8$$
$$\sqrt{81}=9$$
$$\sqrt{16}=4$$
$$\sqrt{100}=10$$
$$\sqrt{121}=11$$
$$\sqrt{144}=12$$
$$\sqrt{169}=13$$

## Geometric Representation

A square root is simply the length of one side of a square. Whatever the area of the square is, the root is the same number that represents the area, enclosed by the radical symbol[2].

For example, the area of this square is 2m²:

This may be pronounced as "two square metres" or "two metres squared"

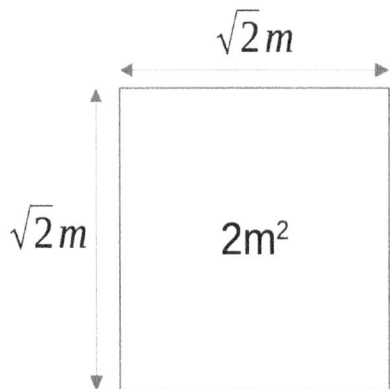

$\sqrt{2}\,m$

$\sqrt{2}\,m$    2m²

---

2    If the area of the square is a perfect square. For example, if the area is 9m², the length of a side is 3m

You may be wondering how we can have a square with the length of its sides being $\sqrt{2}$ ?

After all, we have just seen that $\sqrt{2}$ is a number with an infinite number of decimal places which implies that we can never measure it exactly.

Consider a square with a side length of 1, now draw a diagonal across the square and use Pythagoras' theorem to determine the length of the diagonal.

The result is that the diagonal of this square is $\sqrt{2}$ of whatever units we are using. We can use this fact to construct the square in our previous example as follows:

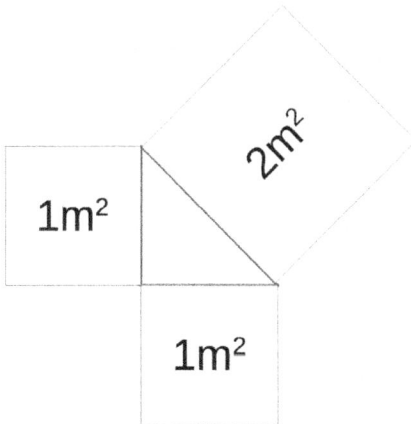

# So What is the point of a Surd?

We have seen so far that the square root of a perfect square is not a surd. We have also defined a surd as the root of a number where the root is irrational.

Surds are results found in a wide range of applications, such as engineering, electronics, physics and many others involving squares and quadratic formulae.

We can show that the square roots of all integers, except perfect squares; are irrational [3]

But what is the point of a surd? We know we can write the square root of 2 as 1.4142 when it is rounded to four decimal places; why not just use a rounded version of the number? Why have an extra symbol $\sqrt{\phantom{x}}$ just for roots?

The answer is accuracy – if you round a number and then multiply it by some value, the bit you rounded off will not be included in the multiplication.

$$1.4142 \times 1.4142 = 1.99996164$$

but

$$\sqrt{2} \times \sqrt{2} = 2$$

If rounded values are used in a series of calculations, these errors increase as more multiplications are carried out. However, if the roots are left in their surd form, no errors are introduced. If an actual decimal value is required, it is best to leave surds in their surd form until all other calculations have been carried out. You should only round values at the last stage of your calculation. This

---

3 Appendix 3 - The Root of any Integer is Irrational except for perfect squares.

can be particularly significant when working with very large or very small numbers, often found in disciplines such as economics, astrophysics and high-energy physics.

**The rounded decimal expansion of a surd can never be as accurate as the actual surd**

# Methods for rationalising Surds

We have seen that a surd is the square root of an integer that is not a perfect square. But, there is a bit more to it than that. Surds may be found in fractions, they could be multiplied together, or added or subtracted to or from integers.

A common requirement for manipulating surds is to reduce them to their simplest form, including manipulating fractions to eliminate surds from the denominator[4]

For example, consider: $\dfrac{3}{\sqrt{2}}$

If we multiply this by $\dfrac{\sqrt{2}}{\sqrt{2}}$ which is equivalent to 1, the result is:

$$\frac{3}{\sqrt{2}} \, x \, \frac{\sqrt{2}}{\sqrt{2}} = \frac{3\sqrt{2}}{\sqrt{2} \, x \, \sqrt{2}} = \frac{3\sqrt{2}}{2}$$

This has removed the surd from the denominator of the solution, or rationalised the denominator We will see more about how to do this in the following sections.

Here are some methods that cover many of the common ways that you may need to rationalise expressions that include surds. First we will list them, then we will look at each one in more detail and see how to use them.

$\sqrt{ab} = \sqrt{a} \, x \, \sqrt{b}$

---

4    Removing surds is known as rationalising because surds are irrational numbers, so getting rid of them results in rational denominators which removes the need to divide by irrational numbers. Scientific calculators often do this automatically.

$$\frac{a}{\sqrt{b}} = \frac{a}{\sqrt{b}} \times \frac{\sqrt{b}}{\sqrt{b}} = a\frac{\sqrt{b}}{b}$$

$$\sqrt{\frac{a}{b}} = \frac{\sqrt{a}}{\sqrt{b}}$$

$$a\sqrt{b} + c\sqrt{b} = (a+c)\sqrt{b}$$
$$note\ that\ a\sqrt{b} + c\sqrt{d} \neq (a+c)\sqrt{b+d}$$

$$a\sqrt{b} \times c\sqrt{d} = ac\sqrt{bd}$$

$$\frac{a}{b+\sqrt{c}} = \frac{a}{b+\sqrt{c}} \times \frac{b-\sqrt{c}}{b-\sqrt{c}} = \frac{ab - a\sqrt{c}}{b^2 - c}$$

$$\sqrt[m]{\sqrt[n]{a}} = \sqrt[mn]{a}$$

The following is not a method, just a useful note:

$$\sqrt[n]{a^n} = a\ if\ n\ is\ odd$$
$$\sqrt[n]{a^n} = |a|\ if\ n\ is\ even$$

## Method 1 The surd of a product is equal to the product of the surds of its factors

$$\sqrt{ab} = \sqrt{a}.\sqrt{b}$$

Consider the rules of indices/powers[5], such as $2^2$ It may help to understand this surd rule by considering them as indices.

A square root can also be written as the number to the power of one half because if we square a root, in power form, then add the indices, we get

$$\sqrt{n} \times \sqrt{n} = n^{\frac{1}{2}} \times n^{\frac{1}{2}} = n^{\frac{1}{2}+\frac{1}{2}} = n^1 = n$$

Similarly, a cube root can be represented in index form as:

$$\sqrt[3]{n} = n^{\frac{1}{3}}$$

In general, a fractional index can be written in surd form as:

$$n^{\frac{a}{b}} = \sqrt[b]{n^a}$$

Consequently, $\sqrt{ab} = \sqrt{a}.\sqrt{b}$

because $a^i \times b^i = (a \times b)^i$ *(see Appendix 4 – Rules for indices)*

$$so\ if\ i = \frac{1}{2}$$

$$(ab)^{\frac{1}{2}} = a^{\frac{1}{2}} \times b^{\frac{1}{2}}$$

_____

5    See Appendix 4 – Rules for indices

Using this relationship, consider the square root of 45

$$\sqrt{45} = \sqrt{9 \times 5} = \sqrt{9} \times \sqrt{5}$$

However, $\sqrt{9} = 3$

So we can rewrite $\sqrt{45}$ as $3\sqrt{5}$

Like the example above, you may well see square roots of large numbers that you can simplify to the square root of a perfect square multiplied by a surd.

One of our goals with all of these rules is to simplify rooted values to minimise the size of the surd and to separate out any integers resulting from the root of perfect squares.

One of the challenges of this technique is to spot the perfect squares so that they can be removed from the radical. For example:

$$\sqrt{24} = \sqrt{4} \times \sqrt{6}$$
$$\text{, but } \sqrt{4} = 2$$
$$so \sqrt{24} = 2\sqrt{6}$$

We may see that $\sqrt{24}$ is also equal to $\sqrt{3}, \sqrt{8}$ but neither 3 nor 8 are perfect squares.

If the 'hidden' perfect square is not obvious, break the number into its prime factors then look for pairs of factors.

For example consider sqrt {72}

$$\sqrt{72} = \sqrt{2} \times \sqrt{2} \times \sqrt{2} \times \sqrt{3} \times \sqrt{3}$$
$$\sqrt{72} = \sqrt{2 \times 2} \times \sqrt{2} \times \sqrt{3 \times 3}$$
$$\sqrt{72} = \sqrt{4} \times \sqrt{2} \times \sqrt{9}$$
$$\sqrt{72} = 2 \times \sqrt{2} \times 3$$
$$\sqrt{72} = 6\sqrt{2}$$

## Exercises – Method 1

Show the following in their simplest form

    1.  $\sqrt{36}$

    2.  $\sqrt{169}$

    3.  $\sqrt{48}$

    4.  $\sqrt{24}$

    5.  $\sqrt{6} \, x \, \sqrt{8}$

# Method 2 Rationalise the denominator when the denominator is a surd or a multiple of surds

$$\frac{a}{\sqrt{b}} = \frac{a}{\sqrt{b}} \; x \; \frac{\sqrt{b}}{\sqrt{b}} = \frac{a\sqrt{b}}{b}$$

In its simplest form, for a fraction with a surd as its denominator, this method multiplies the numerator and the denominator by this same surd. This is the equivalent to multiplying the fraction by one.

The new denominator becomes the original denominator squared which is the radicand of the surd.

A more general solution is shown below.

$$\frac{a}{b\sqrt{c}} = \frac{a}{b\sqrt{c}} \; x \; \frac{\sqrt{c}}{\sqrt{c}} = \frac{a\sqrt{c}}{bc}$$

## Exercises Method 2

Rationalise the denominator in the following:

1. $\dfrac{1}{\sqrt{3}}$

2. $\dfrac{3}{\sqrt{5}}$

3. $\dfrac{4}{\sqrt{8}}$

4. $\dfrac{2}{3\sqrt{6}}$

5. tangent of $x^0$

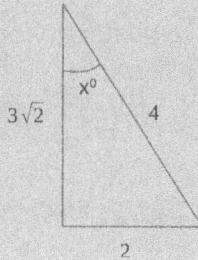

# Method 3 The root of a fraction is equal to the root of the numerator over the root of the denominator

$$\sqrt{\frac{a}{b}} = \frac{\sqrt{a}}{\sqrt{b}}$$

This is a restatement of Index Rule 5 used in Method 1.

Consider $\sqrt{\dfrac{36}{64}}$

You could simplify the fraction first giving $\sqrt{\dfrac{9}{16}}$ and Method 3

The root of a fraction is equal to the root of the numerator over the root of the denominator says this can be rewritten as $\dfrac{\sqrt{9}}{\sqrt{16}}$

which in this example is a perfect square over a perfect square, $\dfrac{3}{4}$

Lets try another that is not a multiple of two perfect squares

$$\sqrt{\frac{36}{40}} = \frac{\sqrt{36}}{\sqrt{4} \, x \, \sqrt{10}} = \frac{6}{2\sqrt{10}} = \frac{3}{\sqrt{10}}$$

As stated at the beginning of this section, it is conventional to get rid of the surd in the denominator whenever we can. In this example, multiply the numerator and the denominator by $\sqrt{10}$

$$\frac{3}{\sqrt{10}} = \frac{3}{\sqrt{10}} \, x \, \frac{\sqrt{10}}{\sqrt{10}} = \frac{3\sqrt{10}}{10} = \frac{3}{10}\sqrt{10} \text{ or } 0.3\sqrt{10}$$

## Exercises – Method 3

Show the following surds in their simplest form

1. $\sqrt{\dfrac{16}{36}}$

2. $\sqrt{\dfrac{5}{36}}$

3. $\sqrt{\dfrac{25}{125}}$

4. $\sqrt{\dfrac{16}{288}}$

## Method 4 Adding two multiples of the same surd is equal to the sum of the multiples multiplied by the surd

$a\sqrt{b}+c\sqrt{b} = (a+c)\sqrt{b}$

What we are doing here is factorising the left-hand side, moving $\sqrt{b}$ outside the brackets.

For example $2\sqrt{2}+3\sqrt{2} = (2+3)\sqrt{2} = 5\sqrt{2}$

*note that* $a\sqrt{b}+c\sqrt{d} \neq (a+c)\sqrt{b+d}$

### Exercises – Method 4

Simplify the following expressions:

1. $3\sqrt{2}+5\sqrt{2}$
2. $7\sqrt{3}+2\sqrt{3}$
3. $6\sqrt{5}-2\sqrt{5}$
4. $6\sqrt{5}-5\sqrt{6}$
5. $7\sqrt{3}-3\sqrt{12}$

## Method 5 The multiple of two surds with coefficients is equal to the multiple of the coefficients multiplied by the root of the multiple of the two surds

$a\sqrt{b} \times c\sqrt{d} = ac\sqrt{bd}$

This is a restatement of the commutative law for multiplication which states that the sum or product of two or more numbers is unaltered if their order is changed.

19

i.e. ab=ba

in this case:

$a\sqrt{b} x c\sqrt{d}$ is equivalent to $a x \sqrt{b} x c x \sqrt{d}$

and

$a x \sqrt{b} x c x \sqrt{d}$ is equivalent to $ac\sqrt{b}\sqrt{d}$

and the Method 1 above shows that this is equivalent to:

$ac\sqrt{bd}$

## Exercises – Method 5

1. $2\sqrt{3} x 3\sqrt{2}$

2. $3\sqrt{2} x 2\sqrt{2}$

3. $3\sqrt{2} x 2\sqrt{6}$

4. $-3\sqrt{6} x 2\sqrt{10}$

5. $7\sqrt{11} x \frac{1}{2}\sqrt{3}$

# Method 6 Use the difference of squares to rationalise the denominator when it contains a sum including a surd

$$\frac{a}{b+\sqrt{c}} = \frac{a}{(b+\sqrt{c})} \ x \ \frac{(b-\sqrt{c})}{(b-\sqrt{c})} = \frac{ab-a\sqrt{c}}{b^2-c}$$

To unravel this process we first need to look at the difference of two squares, for example, if we multiply (a + b) by its conjugate[6]

---

6    The conjugate of a binomial expresion is the same two terms with the ± sign reversed

$$(a+b)(a-b)$$

expand the brackets

$$a^2 - ab + ba - b^2$$

-ab+ba cancels out, leaving a²- b² which is the difference of two squares.

A geometric way of describing this is the that difference of the area of two squares is equal to the area of one square minus the area of the other, as shown here

The area of the bigger square is a² and the area of the smaller square is b²

The area of the difference is the sum of the two rectangles, i.e. a(a-b)+b(a-b)

Factorize by (a-b) gives

(a-b)(a+b)

and this area is the difference in the area of the two squares a²-b²

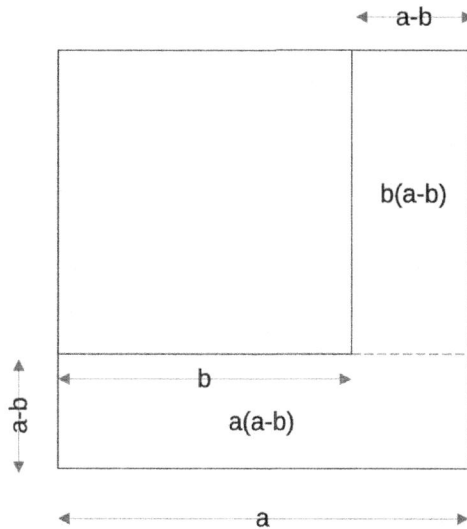

Consequently, if we have an expression with a square root in the denominator, plus *(or minus)* another number, we can multiple it by the original denominator with the ± sign reversed, the resulting denominator will not have a surd in it.

Here is an example:

$$\frac{3}{2+\sqrt{2}}$$

multiply numerator and denominator by $2-\sqrt{2}$

$$\frac{3(2-\sqrt{2})}{(2+\sqrt{2})(2-\sqrt{2})}$$

expand the brackets and simplify:

$$\frac{6-3\sqrt{2}}{4-2\sqrt{2}+2\sqrt{2}-2} = \frac{6-3\sqrt{2}}{2} = 3-\frac{3\sqrt{2}}{2}$$

As you can see above, the denominator, after expanding the brackets, is the original number squared minus the surd squared, and the surd squared is the surd's radicand which can be added or subtracted from the other number.

## Exercises – Method 6

Rationalise the denominator and simplify where possible:

1. $\dfrac{6}{3\sqrt{2}}$

2. $\dfrac{1}{\sqrt{5}-2}$

3. $\dfrac{6}{\sqrt{7}+1}$

4. $\dfrac{\sqrt{3}+2}{2-\sqrt{3}}$

5. $\dfrac{\sqrt{5}-2}{\sqrt{2}\left(4+\sqrt{3}\right)}$

# Method 7 - The root of a root is equal to the root of the product of their indices

$$\sqrt[m]{\sqrt[n]{a}}=\sqrt[mn]{a}$$

This can be illustrated using Rule 4 of the Rules of Indices *(see page 38)* $(a^m)^n = a^{mn}$

we can rewrite $\sqrt[m]{\sqrt[(n)]{a}}$ as $\left(a^{\frac{1}{n}}\right)^{\frac{1}{m}}$

Indices rule 4 says this is equal to $a^{\frac{1}{m}\times\frac{1}{n}}$

which is equal to $a^{\frac{1}{mn}}$

For example, to simplify $\sqrt[20]{1024}$

First change it to $\sqrt[5]{\sqrt[4]{1024}}$

23

Now simplify the inner root $\sqrt[4]{1024} = \sqrt[4]{(2^{10})} = (2^{10})^{\frac{1}{4}} = 2^{\frac{10}{4}} = 2^{\frac{5}{2}}$

Now simplify the outer root $\sqrt[5]{2^{\frac{5}{2}}} = (2^{\frac{5}{2}})^{\frac{1}{5}} = 2^{\frac{5}{10}} = 2^{\frac{1}{2}}$

$$= \sqrt{2}$$

# Exercises - Method 7

Simplify the following expressions giving your answers as either integers or surds:

1. $\sqrt[2]{\sqrt[2]{16}}$

2. $\sqrt[3]{\sqrt[2]{64}}$

3. $\sqrt[4]{\sqrt[2]{81}}$

4. $\sqrt[3]{\sqrt[3]{729}}$

5. $\sqrt[6]{\sqrt[3]{4096}}$

# Calculating the Approximate Value of Surds

There are several methods you can use to calculate the approximate value of a surd, depending on the accuracy required and the calculation methods available such as pencil and paper, calculator or computer. In some cases, the first step is to decide on an initial value.

For example, if you want a quick approximation and you have a calculator.

Consider the square root of 83.

First find the perfect squares that are closest to 83. In this example, 81 and 100 – the square root must be greater than 9 and less than 10

Next, subtract 81 from 83 giving 2, then 81 from 100, giving 19.

83 is $\frac{2}{19}$ of the proportion of the difference between 81 and 100

Try $\left(9\frac{2}{19}\right)^2$ as the first estimate. This results in 82.91 to 2dp, i.e. 0.09 less than the radicand.

## The Long Division Method

This method can be used to calculate a root with just a pencil and paper. It can be used for any number, including those with multiple decimal places.

Step 1 is to separate the digits into pairs, moving in both direction, starting from the decimal point. for example:

540,577.8576will become:

54 | 05 | 77 | . | 85 | 76

Step 2 is to take the left-most pair and find the nearest perfect square that is less than it; in this case $7^2=49$

This is the first digit of the result.

Write result down above the first pair.

Subtract the squared value from the first pair.

The following steps are now repeated until the required accuracy is reached, or a perfect square is found.

Step 3 Bring down the next pair, appended to the result of the subtraction. This provides a dividend.

```
        7
 ─────────────────────────────
 54 | 05 | 77 | . | 85 | 76
 49
     5    05
```

i.e 54-49=5 next pair is 05 so the dividend is 505

Step 4 is to double the result, then add a digit to the end of the doubled result to find a divisor that, when multiplied by this same digit, is the highest multiple that is less than the dividend.

```
    7
54 | 05 | 77 | . | 85 | 76
49
 5   05
```

2 x current result = 14

505 ÷ (14? x ?) must be less than 505

In this example, the divisor will be 14? such that 14? x ? Is the highest multiple of 14? that is less than 505. In this case 144 x 4 = 576 which is greater than 505 so our divisor is 143.

The 3 you added to the end of the divisor is the next digit of the result so write it down above the pair.

```
    7   3
54 | 05 | 77 | . | 85 | 76
```

Repeat steps 3 and 4 until you either find a perfect square or the required number of digits.

The complete set of steps are shown on the next page.

```
                            7
                 ─────────────────────────────
                 54 │ 05 │ 77 │ . │ 85 │ 76
                 49
                  5    05
```

7 x 2 = 14                       505
14? x ?                         -429
14<u>3</u> x <u>3</u> = 429        76    77

```
                  7     3            .
                 ─────────────────────────────
                 54 │ 05 │ 77 │ . │ 85 │ 76
```

73 x 2 = 146                      7677
1,46? x ?                        -7325
1,46<u>5</u> x <u>5</u> = 7,325      352         85

```
                  7     3     5
                 ─────────────────────────────
                 54 │ 05 │ 77 │ . │ 85 │ 76
```

735 * 2 = 1470                        35285
14,70? x ?                           -29404
14,70<u>2</u> x <u>2</u> = 29,404        5881    76

```
                  7     3     5    .     2
                 ─────────────────────────────
                 54 │ 05 │ 77 │ . │ 85 │ 76
```

7352 x 2 = 14704                        588176
147,04? x ?                            -588176
147,04<u>4</u> x <u>4</u>=588,176            0

```
                  7     3     5    .     2     4
                 ─────────────────────────────
                 54 │ 05 │ 77 │ . │ 85 │ 76
```

# Herron's Method

Although, three thousand years ago, the Babylonians appeared to have had a method for calculating $\sqrt{2}$ to three decimal places[7], the first known algorithm was identified by the Greek Heron in the first century AD.

The algorithm is based on making an initial estimate, then finding the mean of the initial estimate and the radicand divided by the esmate.

If the estimate is higher than the root, the radicand divided by the estimate must be less than the root, so the mean of these two numbers will be closer to the actual root than the first estimate.

If the estimate is lower than the root, the radicand divided by the estimate must be greater than root, so the mean of these two numbers must also be closer to the actual root than the first estimate.

The resulting mean becomes the new estimate and this process can be repeated until the required accuracy is obtained.

---

7   https://en.wikipedia.org/wiki/
    Methods_of_computing_square_roots#Heron's_method

This algorithm, for $\sqrt{S}$ where the initial guess is $x_n$ and can be expressed as:

$$x_{n+1} = \frac{1}{2}\left(x_n + \frac{S}{x_n}\right)$$

$x_{n+1}$ then replaces $x_n$ and the process is repeated.

This algorithm is suitable for programming problems and a simple Python program is given in Appendix 5 – Herron's Method in Python

# Appendices

# Appendix 1 - The Square Root of 2 is Irrational

We will prove that the square root of 2 is irrational. This means that the square root of two can not be represented by a fraction.

We will prove this by contradiction, meaning that we will make a statement that is the opposite of what we want to prove, and then show that the statement is false,

First, state that the square root of 2 is a fraction in its simplest form and then show that this is not true.

$\sqrt{2} = \frac{p}{q}$ where $\frac{p}{q}$ is in it's simplest form, meaning that $p$ and $q$ do not share any common factors other than 1. Another way of saying this is that p and q are coprime, meaning that if $p$ and $q$ are reduced to their prime factors, they do not have any in common.

For example $\frac{15}{22}$ meets this criterion as 15 and 22 are coprime:

$$15 = 3 \times 5$$

$$22 = 2 \times 11$$

This means $\frac{15}{22}$ is in its simplest form.

Now we square both sides of our original equation:

$$\left(\sqrt{2}\right)^2 = \left(\frac{p}{q}\right)^2$$

$$2 = \frac{p^2}{q^2}$$

Now multiply both sides by $q^2$ giving

$$2q^2 = p^2$$

Any number multiplied by 2 is an even number

$p^2$ is equal to $p$ x $p$. For $p^2$ to be even, $p$ must also be even *(see Appendix 2 - If the Square of a number is even, the number must also be even)* therefore $p^2$ is even.

Because $p$ is even, we can replace it with $(2k)^2$

Our equation now becomes:

$$2q^2 = (2k)^2$$

Expand the right hand side bracket giving

$$2q^2 = 4k^2$$

Divide both sides by 2 giving

$$q^2 = 2k^2$$

$q^2$ must be even because it is equal to something multiplied by 2. It is also equal to $q$ x $q$. For this to be even, $q$ must also be even. Therefore both $p$ and $q$ are even numbers which means they must both be divisible by 2. But we started off by stating that $\sqrt{2} = \dfrac{p}{q}$ where $\dfrac{p}{q}$ is in it's simplest form but this has been contradicted by our initial assumption.

# Appendix 2 - If the Square of a number is even, the number must also be even

An even number is any integer multiplied by 2, 2n in general

An odd number is either $2n + 1$ or $2n - 1$

We can see that multiplying an even number by either an odd or an even number will always result in an even number.
For two integers, i and j, multiply:

even by even $= 2i.2j = 4ij = 2(2ij) = 2(\text{some number})$

even by odd $\quad\quad = 2i(2j + 1) = 4ij + 2i = 2(2ij + I)$

$\quad\quad\quad\quad\quad = 2(\text{some number})$

or

$\quad\quad\quad\quad\quad = 2i \times (2j - 1) = 4ij - 2i = 2(2ij - I)$

$\quad\quad\quad\quad\quad = 2(\text{some number})$

An odd number multiplied by an odd number is always odd.
The square of a number is the number multiplied by itself.

odd by odd $\quad\quad = (2i + 1)(2j + 1) = 4ij + 2i + 2j + 1$

$\quad\quad\quad\quad\quad = 2(2ij + i + j) + 1 = 2(\text{some number}) + 1$

or

$\quad\quad\quad\quad\quad = (2i - 1)(2j - 1) = 4ij - 2i - 2j + 1$

$\quad\quad\quad\quad\quad = 2(2ij - i - j) + 1 = 2(\text{some number}) + 1$

Consequently, if we square an odd number the square must also be odd *(odd x odd = odd)*.

If the number is even, the square must be even *(even x even = even)*.

*(even x odd)* is not an option as we are multiplying the number by itself.

Therefore, if $a^2$ is even, $a$ must also be even.

Similarly, if $a^2$ is odd, $a$ must also be odd.

# Appendix 3 - The Root of any Integer is Irrational except for perfect squares.

We will use a proof by contradiction.

Assume $\sqrt{n} = \dfrac{p}{q}$ where n is an integer and p and q are coprime integers, i.e. the fraction $\dfrac{p}{q}$ is in its simplest form which would mean that $\sqrt{n}$ is a rational number if our assumption is correct.

We will prove that $\sqrt{n} = \dfrac{p}{q}$ is invalid meaning that $\sqrt{n}$ must be an irrational number.

p and q are each the product of a number of primes; this is the fundamental theorem of arithmetic.

$$p = a_1 a_2 a_3 \dots a_i$$
$$q = b_1 b_2 b_3 \dots b_i$$

As p and q are coprime, $a \neq b$ for all values of $a$ and $b$.

Square both sides

$$n = \dfrac{p^2}{q^2}$$

Next, what if q is one?

If q is one, so is $q^2$ which means that n = $p^2$, but this makes n the square of p and we are not including perfect squares in this definition.

If q is not one, $\dfrac{p^2}{q^2}$ must be a fraction, not an integer, as p and q do not share any common factors. Therefore, n must also be a fraction.

But, our assumption was that n was an integer which is a contradiction.

Therefore $\sqrt{n}$ can not be represented by a fraction and is therefore irrational.

This same logic can be applied to any root of n.

# Appendix 4 – Rules for indices

**Rule 1:** $a^m \times a^n = a^{m+n}$

Consider $a^m \times a^n$

This is equal to ($a$ x $a$ x $a$...m times) x ($a$ x $a$ x $a$...n times)

because:

  ($a$ x $a$ x $a$...m times) x $a = a^{m+1}$

  ($a$ x $a$ x $a$...m times) x $a$ x $a = a^{m+2}$

  ($a$ x $a$ x $a$...m times) x ($a$ x $a$ x $a$...n times)

  which is $a^{m+n}$

e.g. $2^3 \times 2^2 = (2 \times 2 \times 2) \times (2 \times 2) = 2^5$

**Rule 2:** $\dfrac{a^m}{a^n} = a^{m-n}$

$$\frac{(a\,x\,a\,x\,a\ldots \text{ m times })}{(a\,x\,a\,x\,a\ldots \text{ n times })}$$

If m > n, each $a$ in denominator the will cancel with one $a$ in the numerator, leaving (m – n) x $a$ in the numerator

e.g.    $\dfrac{2^4}{2^2} = \dfrac{16}{4} = 4$

and    $\dfrac{2^4}{2^2} = 2^{4-2} = 2^2 = 4$

**Rule 3: $a^0 = 1$**

Following from the second rule

$$\frac{a^m}{a^m} = a^{m-m} = a^0$$

but $\dfrac{a^m}{a^m} = 1$ therefore $a^0 = 1$

**Rule 4: $(a^m)^n = a^{mn}$**

$(a^m)^n = (a^m) \times (a^m) \times (a^m)\ldots$ *(n times)*

$(a^m) \times (a^m) = a^{2m}$ - *from Law 2*

$(a^m) \times (a^m) \times (a^m) = a^{3m}$

$(a^m) \times (a^m) \times (a^m)\ldots$ *(n times)* $= a^{nm}$

**Rule 5: $a^m \times b^m = (ab)^m$**

Also, to multiply two different numbers with the same index, we can first multiply the numbers together, then raise the result to the index.

$a^m \times b^m = (ab)^m$ *(factorise by taking the power of $^m$ outside the brackets)*

e.g. $2^3 \times 3^3 = 8 \times 27 = 216$ which is the same as:

$(2 \times 3)^3 = 6^3 = 216$

Similarly, $\dfrac{a^m}{b^m} = \left(\dfrac{a}{b}\right)^m$

**Rule 6:** $a^{-m} = \dfrac{1}{a^m}$

Consider $\dfrac{a^m}{a^{m+1}}$

Each $a$ in the numerator will cancel out with each $a$ in the denominator, leaving on extra $a$ in the denominator, i.e. $\dfrac{1}{a}$

Additionally, Rule 2 says that $\dfrac{a^m}{a^n} = a^{m-n}$

and in this case n = (m+1)

so we will have an index of m − (m + 1) which equals -1

Therefore $a^{-1} = \dfrac{1}{a}$

**Rule 7:** $a^{\frac{1}{2}} = \sqrt{a}$

From Rule 1: $a^m \times a^n = a^{m+n}$

$a^{\frac{1}{2}} \times a^{\frac{1}{2}} = a^1 = a$

or

$\sqrt{a} \times \sqrt{a} = a$

# Appendix 5 – Herron's Method in Python

```
1   #!/usr/bin/python
2   i=0
3   j=0
4   #while(j <4):
5
6   while True:
7       radicandString= input("Radicand ")
8       if radicandString == "" :
9           break
10      radicand=float(0)
11      try:
12          radicand=float(radicandString)
13      except ValueError:
14          break
15      newEstimate=float(1)
16      estimate=float(2)
17      while(estimate!=newEstimate):
18          estimate=newEstimate
19          newEstimate=(estimate+radicand/estimate)/2
20          i+=1
21      print("Root is ",newEstimate, " in ",i, " iterations")
22      j+=1
23      False
24   print (j," roots calculated")
```

```
Radicand 4
Root is  2.0  in  7  iterations
Radicand 9
Root is  3.0  in  14  iterations
Radicand 2
Root is  1.414213562373095  in  20  iterations
Radicand 37.69
Root is  6.139218191268331  in  28  iterations
Radicand
4  roots calculated
```

This is a listing of a simple Python 3 program that uses Herron;s method to calculate the square root of a number that is entered when the program is run.

The outer while-loop repeats until there is a break following a non-numeric entry.

The inner while loop repeats until there have been two successive results that are equal. This equality does not imply necessarily that a square root has been found *(i.e. that the radicand is a perfect square)* rather, it is the result of the limited accuracy of computer calculations.

The actual calculation is on line 10.

The results are shown below the program.

# Appendix 6 - Roots and Zeros

There are several similar terms relating to the use of the word root.

A square root, the main subject of this book, is a number, when multiplied by itself, results in the number that it is the square root of.J

A more general term is the $n^{th}$ root, a number, when multiplied by itself n times results in the number it is the $n^{th}$ root of. For example, 3 is the $4^{th}$ root of 81.

A related use of root is a value that makes an equation true. for example, for the equation $x^2 - 4 = 0$, the root is either 2 or -2.

We sometimes need to find the roots of a more complex quadratic equation, for example:

$$f(x) = x^2 - 5x + 6$$

If f(x) is zero, this can be factorised to $f(0) = (x - 2)(x - 3)$

Therefor for x = 2 or 3, the equation is true so 2 and 3 are the roots of this equation.

Here is a graph of that equation. As you can see the roots occur when the function crosses the x axis

This is why roots of functions are also known as zeros of the function.

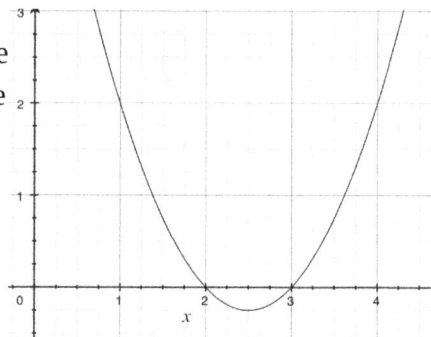

A well known use of the term "zeros" was a statement made by Bernard Rienmann in 1859.

A translation of Rienmann's Hypothesis is

"The non-trivial zeros of the Rienmann Zeta Function all have a real part of one half

The Rienmann Zeta Function is a series of complex numbers and, although few if any people understood his statement until several years after his death, it has become recognised as a vital problem, relating to the distribution of prime numbers, either to be proved true or false. Some of the best mathematical minds have wrestled with this problem for over 150 years and a prize of $1,000,000 has been offered for a solution since the year 2000.

An excellent book about Bernard Rienmann, which inclued a concise history of maths, is "A Prime Obsession" by John Derbyshire.

# Answers

## Page 16 Method 1

1    6

2    13

3    $4\sqrt{3}$

4    $2\sqrt{6}$

5    $4\sqrt{3}$

## Page 17 Method 2

1    $\dfrac{\sqrt{3}}{3}$

2    $\dfrac{3\sqrt{5}}{5}$

3    $\sqrt{2}$

4    $\dfrac{\sqrt{6}}{9}$

5    $\dfrac{\sqrt{2}}{3}$

## Page 18 Method 3

1    $\dfrac{2}{3}$

2    $\dfrac{\sqrt{5}}{6}$

3    $\dfrac{\sqrt{5}}{5}$

4    $\dfrac{\sqrt{2}}{12}$

## Page 19 Method 4

1    $8\sqrt{2}$

2    $9\sqrt{3}$

3    $4\sqrt{5}$

4    already in its simplest form, the surds have no common factors

5    $\sqrt{3}$

## Page 20 Method 5

1    $6\sqrt{6}$

2    12

3    $12\sqrt{3}$

4    $12\sqrt{15}$

5    $\dfrac{7}{2}\sqrt{33}$

## Page 23 Method 6

1    $\sqrt{2}$

2    $\sqrt{5}+2$

3    $\sqrt{7}-1$

4      $7 + 4\sqrt{3}$

5

$$\frac{4\sqrt{10} - \sqrt{30} - 8\sqrt{2} + 2\sqrt{6}}{26}$$

## 24 Method 7

1      2

2      2

3      $\sqrt{3}$

4      $\sqrt[3]{9}$

5      $\sqrt[3]{4}$

Printed in Dunstable, United Kingdom